Reality Technologies - The Conceptual Framework

Table of Contents

Preface..6
1.0 Introduction.................................... 10
2.0 Virtual Reality................................16
3.0 Augmented Reality.......................... 43
4.0 Mixed Reality.....................................54
5.0 Telecommunication Infrastructure
 Challenges.....................................56
Referred Standards....................................67
Recommended Reading...............................72
Key Terms...74
Summary...76
References..78
Check Your Learning................................80
About the Author.......................................85

List of Figures

Figure 1A – Reality Technology Spectrum............13

Figure 2 – Understanding Reality Technologies......14

Figure 3 – Virtual Ecosystem...........................21

Figure 4 – Virtual Environment with 6DoF.........32

Figure 5 – Understanding the Field of Vision........40

Figure 6 – Telecom Technologies – New Developments..61

Figure 7 – Evolved Telecom Ecosystem................67

List of Tables

Table 1 – Levels of Immersion..........................24

Table 2 – Virtual Reality Sensory Perceptions......25

Table 3 – Virtual Reality Hardware...................27

Table 4 – FOV Values...................................41

Table 5 – Bandwidth Requirements for Common Reality Applications......................................58

Table 6 – Latency and Bandwidth requirements of new age applications.....................................59

Table 7 – Structural Changes in Telecom Network to support Reality Technologies..........................63

Dedicated to my family….

PREFACE

The world as we know of today and the way we live our lives will undergo a sea change in the next five years. There are a whole gamut of innovative technologies, straddling virtually every sphere of the vast industrial spectrum, that are going to force mankind to adopt new creative, innovative and technologically driven practices in our day-to-day lives.

It is forecasted that the global demand for Reality hardware will exceed over 110 million units by the year 2020 from the current levels of 9.6 million units. This figure amounts to over 200% growth in the reality hardware market as compared to 2016. The reality application (software) market is also expected to increase by over 100% as compared to the year 2016. The bulk of the demand, as it stands today, is for the display units that includes smart phone based seamless viewers and head mounted display (standalone & tethered) units. The reality industry has crossed the threshold that signifies its emergence as a game changer in the near future.

The reality technologies provide for an enhanced user experience, through the inclusion of virtual objects within real and/or virtual environments, which simulates various real life scenarios and finds application in almost all the industrial sectors. 360°videos provide users with the ability to participate in an immersive experience, over unlimited range of scenarios, that impact virtually all industrial sectors from healthcare (education at medical schools, consultation, treatments and even surgical procedures), media & advertising, manufacturing, retail, education, real estate, entertainment and military. The reality technologies encompass Augmented Reality (real environment and virtual objects),

Augment Virtuality (virtual environment and real objects), virtual reality (virtual environment and virtual objects) and hybrid or mixed (combination of reality technologies).

However the limitations of the current telecom infrastructure prevents the wide spread assimilation of the reality technologies and provide users only a glimpse of the virtually endless applications of these immersive technologies. The reality applications require a large amount of bandwidth to deliver high resolution 360° videos, typically in the range of 25 to 750Mbps with low latency levels of less than 20ms (motion to photon latency) with certain applications (like autonomous driving)requiring latency of less than 10ms to avoid motion sickness. These requirements necessitate a revamped access network architecture that can provide 1Gbps connectivity to the end users.

Further the access networks must adopt switching technology that can support the stringent latency requirements required by reality applications. The overall network architecture should provide for integration of data centers, at the network edge, to support of reality content distribution (to ensure low latency levels.

The exponential increase in the bandwidth envisaged at the access layer, mandates a relook at the core or backbone telecom network architecture, with a view of ensure sufficient capacity, reduced latency and improved availability through enhanced resilience. An agile management framework is also required to ensure real time FCAPS support.

Key Learnings

- Understanding the essence of reality technologies
    - *Augmented Reality*
    - *Augmented Virtuality*
    - *Mixed Reality*
    - *Virtual Reality*
- Describe the hardware and software architecture of reality systems
- Understand the key requirements and challenges for upgradation of telecom infrastructure to support reality technologies

Happy Reading!!!

Sudhir Warier

UNDERSTANDING REALITY TECHNOLOGIES

1.0   Introduction

The Cambridge dictionary [1] defines the noun 'realty' as "the state of things as they are, rather than as they are imagined to be". The noun 'perception' is defined as "a belief or opinion, often held by many people and based on how things seem". The adjective 'virtual' is defined as "almost a particular thing or quality". The noun 'virtual reality' is defined as "A set of images and sounds, produced by a computer, that seem to represent a place or a situation that a person can take part in".

The visual perception, in human beings, provide them the ability to construct three dimensional visual images of their immediate physical environment, based on the processing of the light signals by the brain, augment by the other sensory mechanisms like sense and smell.

The replacement of the sensory experiences by artificially reproduced information, through reality technology, causes the brain to perceive them as 'reality'. The realty technologies can be classified as illustrated in figure 1A and1B. Figure1.A presents a 'birds-eye' view of the entire reality technology spectrum while 1.B presents the three primary technologies and their key differences.

Reality technologies provides an environment (real and/or virtual) and a set of objects (real and/or virtual) with whom the user will engage, based on the nature of the engagement required. This concept is illustrated in the figure 2.

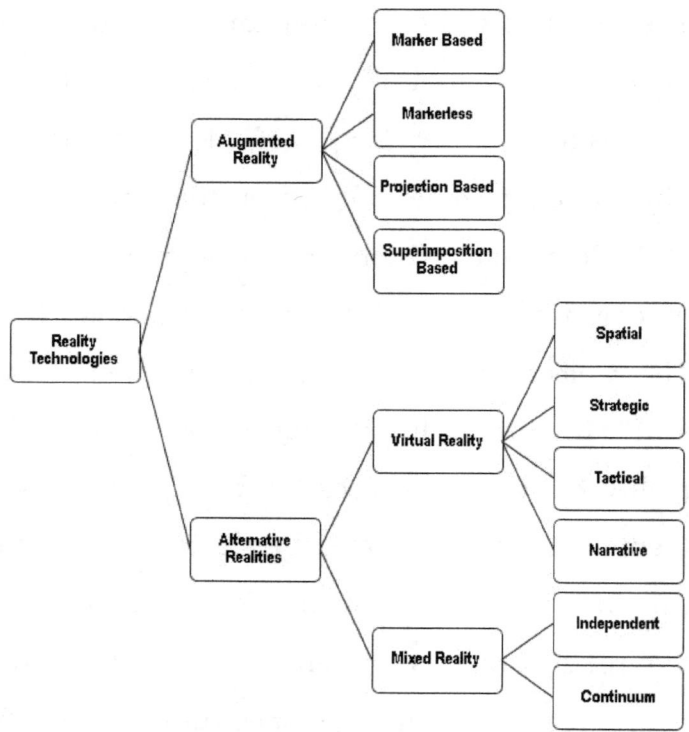

Figure 1A – Reality Technology Spectrum

As is the case with all emerging technologies, the reality technology landscape, currently looks a bit complicated and confusing.

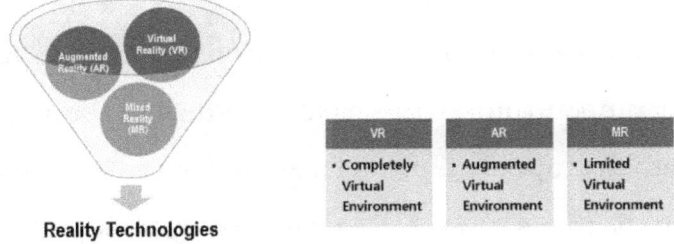

Figure 1B – Primary Reality Technologies

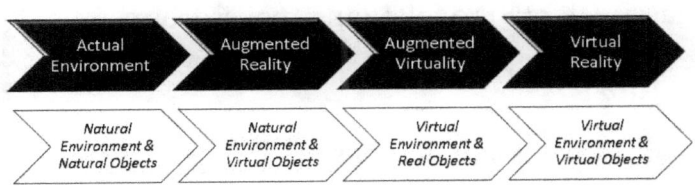

Figure 2 – Understanding Reality Technologies

There are different strategies, technologies and applications that straddle the landscape and a more organized ecosystem is expected to emerge in the near future. The virtual reality landscape consists of three primary and inter-related building blocks that include the applications, supported platforms and the infrastructure required to support the applications.

The reality applications straddle different industrial domains ranging from aerospace to manufacturing to service industry besides military applications. They find wide ranging applications for learning& development interventions.

The platforms for hosting the application can be classified into the following major groups:

1. Capturing
2. Processing Engines
3. Distribution

The capturing mechanisms include the following major components:

1. Light Field Mapping - Includes 3D computer graphics and artificial intelligence (AI) for scene analysis
2. 360^0 Video Capture - Specialized cameras (omnidirectional cameras) with associated

accessories like jigs (requires rectangular videos displayed on spheres)

3. Stereoscopic Video Capture – Specialized cameras for wide-angle stereoscopic video

The processing engines consist of software tools to create, edit and render reality content. These include reality frameworks and engines (augmented, mixed and virtual reality) image processing, stitching software (stereophonic videos) besides compression technologies to handle the huge volumes of data generated.

Distribution platforms provide users a method of accessing virtual reality content through download and/or streaming. The platforms may be based on open architecture or based on specific (and often proprietary) hardware. The examples include Head Mounted Displays (HMD) and/or projection technologies described later in this chapter.

2.0 Virtual Reality

As the name suggests Virtual Reality (VR) [2] refers to an interactive system (hardware and software) that is capable of providing a realistic simulation of an environment to a user, who perceives it as a real life environment. The adjective 'immersive' can be described as "seeming to surround the audience, player, etc. so that they feel completely involved in something". The term can be aptly used to define VR which provides humans with the possibility of having an immersive experience of a range of environments using a combination of hardware (wearable devices) and software (including content). VR facilitates the recreation of an unlimited range of visualization of objects, environments and views that are perceived to be real, at the place and timing desired by the user.

The three dimensional (3D) interactive virtual environment created by the VR system creates a perception of reality for the end user.

The brain believes that it is interacting with a real environment and creates a realistic sensory experience that is defined by the term "immersive" and the process as "Total Immersion". The virtual environment may mimic real life settings or events including 'visiting tourist attractions', taking part in a 'war' or playing a favourite sport.

2.1 Categories

On the basis of the current evolution there can be three categories of VR systems. The categories are based on the levels of immersive experience offered and the related application experience. The categories include:

1. *Fully-Immersive VR Systems*

Fully-Immersive VR systems (VR-FI) provides a realistic environment using a combination of hardware that includes head-mounted displays (HMD), motion sensors that deliver high resolution specially shot content with a wide field of view (80º and higher) high refresh rates and contrast, that simulates the key sensory organs and provide a fully immersive experience.

Example 1

A virtual tour of the Taj Mahal, India

2. *Semi-Immersive Virtual Reality*

As the name implies, Semi Immersive Virtual Reality (VR-SI) systems provide a partially immersive system using high resolution projectors and/or smart televisions driven by high end graphical systems to simulate a real environment. The environment mimics a real life visualization that provides a semi-realistic experience to the end user.

Example 2

Driving Simulators and Fight Simulators

3. ***Non-Immersive Virtual Reality***

Non-immersive Virtual Reality (VR-NI) systems provide a 3-D virtual rendering of a real-life environment. The environment however does not engage all the human senses and the user is aware of his immediate surroundings, reducing the perception of reality. A VR-NI system consists of a high resolution monitor that powered by conventional desktop computing systems that provides simulation of a 3D environment.

Example.3

3D Games

2.2 Virtual Ecosystem

The primary component of the virtual ecosystem is the environment that the user will interact with.

The level and nature of these interactions and their depth (sensory perceptions) form the remaining components of the ecosystem. A virtual ecosystem is illustrated in the figure 3:

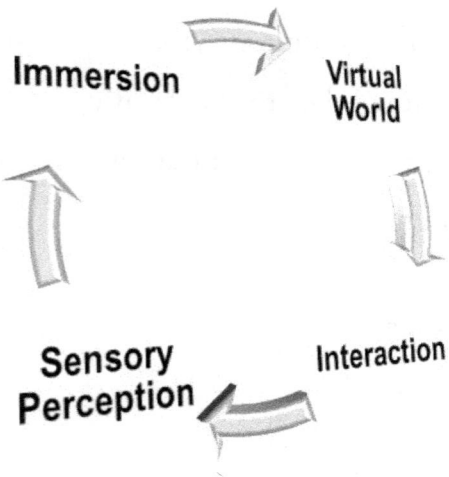

Figure 3 – Virtual Ecosystem

a. *Virtual World*

A virtual world refers to the multi-dimensional environment, realized through a suitable medium, through which the user interacts to observe and/or perform specific activities. The virtual environment responds to the user inputs (movements) in a pre-programmed sequence resulting in changing virtual perspectives (views) mimicking real-life environments and corresponding user interaction.

b. *Interaction/Collaboration*

The nature and/or level of interaction or collaboration between the user and the virtual environment is a crucial factor that determines the success of a VR experience. The nature/level of interaction must be a replica of the interaction in the real environment and the response to user action must resemble the real life responses for a user to get fully immersed with the VR environments.

VR systems must be capable of accurately tracking the head movements (since it is directly related to the field of vision) and/or other body parts (dependent on the environment being simulated). The human brain can quickly perceive latency of responses and directly affect the level of user immersion with the VR system.

c. **Immersion**

Virtual reality can also be defined as the physical interaction of a user in a non-physical (virtual) environment [3].

The virtual environment and its interaction with the user, physically along with the associated mental faculties, makes the brain provide a perception of reality as a result of which the user gets immersed (at varying levels, dependent on the VR system) in the visualization representing the virtual environment.

Virtual reality systems can also be categorized based on the levels of immersion, as perceived by the user. The concept is outlined in table 1:

Table 1 – Levels of Immersion

S.N	Immersion Category	Alternate Name	Description
1	Spatial	-	Perception of reality in virtual environment
2	Strategic	Cognitive Immersion	High levels of mental engagement
3	Tactical	Sensory	Engage in

		Motor Immersion	performing
4	Narrative	Emotional Immersion	Engage with the narration or visuals being presented

d. Sensory Perception

For a fully immersive experience the VR system must be capable of replicating the following 5 senses as presented in table 2:

Table 2 – Virtual Reality Sensory Perceptions

S.N	Sense	Description	Method
1	Visual	Seeing	3D Panoramic Display
2	Auditory	Hearing	Surround Sound

3	Olfactory	Smell	Aromatic Diffuser
4	Gustation	Taste	Bone conduction transducers (coupled with aromatic effects)
5	Tactile	Touch	Haptics (Force)

2.3 Hardware Architecture & Components

Virtual reality hardware plays an important role in creating the virtual environment based on the desired level of immersion. There are different components used based on the sensory perceptions that need to be created. The table 3 lists some of the major hardware components used in VR systems [4]:

Table 3 – Virtual Reality Hardware Components

S.N	Component Type	Hardware
1	Display	Head mounted display (with or without smartphones) [Includes Google cardboard, slot-in devices, HMD, Sleek HMD and imperceptible devices(future)]
		Projection Rooms (CAVE Display)
		PC/Smart Phone (VR display on non-VR systems)
2	Input Devices	Control Buttons
		Controller Wands

		Data Gloves
		Force Balls/ Tracking Balls
		Joysticks
		Motion Platforms
		Motion Trackers/Bodysuits
		Trackpads
		Treadmills
3	Sensors	Magnetometers,
		Accelerometers
		Gyroscopes
4	Tracking	Head Tracking
		Motion Tracking
		Eye Tracking
5	Acoustic	High fidelity 3D surround sound systems

6	Olfactory	Aromatic Diffusers
7	Processors	Input Processor
		Simulation Processor
		Rendering Processor

a. Display

Virtual reality content is presented within an interactive 3D environment that is presented commonly within a head mounted display (HMD). It can also be projected using specialized projectors within projection rooms. VR content can also be presented through non VR systems including personal computing devices including smartphones.

The content requires a processing engine backed by significant computing power (based on the magnitude of immersion).

The processing can be done using personal computers and/or smart phones.

The HMD (using Liquid Crystal Displays (LCD) or Organic Light-Emitting Diode (OLED)) is mounted in front of the user and provides a full field view of the VR content being displayed. The HMD may use a smartphone or provide interfaces like HDMI to connect a suitable engine to process and display the content. An integrated (or separate) headset provides the sound input (3D surround or otherwise) to provide the acoustic input needed for a truly immersive experience.

There may be special configurable lenses incorporated in between the user eyes and the display that facilitates the reshaping of the picture to account for the stereoscopic view of the real world provided by both the human eyes.

b. Input Devices

Input devices are required to provide the users with capability to interact and navigate through the virtual environment. They are mandatory to provide users a complete immersive experience with the VR system.

c. Sensors

For providing a real life simulation in a virtual environment six-degrees-of-freedom (6DoF) is required for a body in a three dimensional space. These include:

i. Surge (Roll) – Forward/Backward movements along X-Axis

ii. Heave (Yaw) – Up/Down Movements along Z-Axis

iii. Sway (Pitch) – Left/Right Movements along Y-Axis

The concept is illustrated in figure 4. A virtual reality headset includes three basic sensors that provide the users with the orientation in a three dimensional environment.

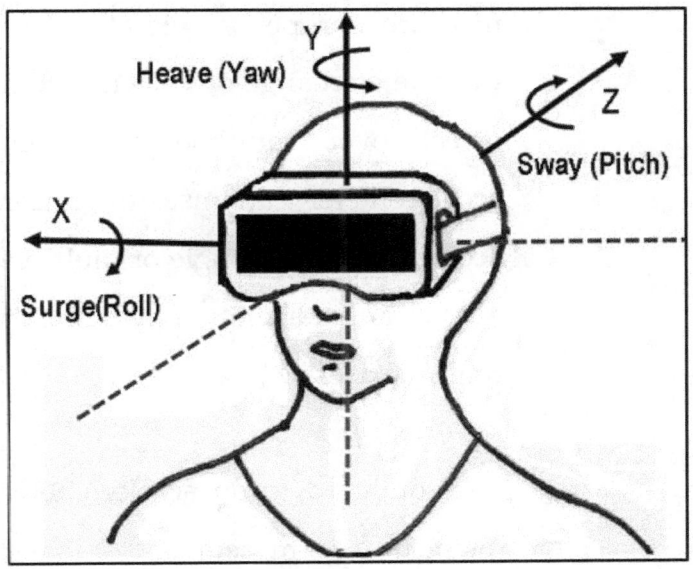

Figure 4 – Virtual Environment with 6DoF

The sensors include:

i. Magnetometers

Magnetometers acts as a compass and provides the VR device with the capability to estimate its direction with respect to the magnetic north.

ii. *Accelerometers*

An accelerometer provides the capability to the VR device to measure 'proper acceleration' or the rate of change of velocity of an object over multiple axes. An accelerometer can measure linear acceleration over single or multiple axes or angular motion over one or multiple axes.

iii. *Gyroscopes*

As opposed to an accelerometer which measure linear acceleration (mV/g), a gyroscope measures angular velocity (mV/deg/s). It is used as a navigational aid.

d. Tracking Mechanisms

Virtual Reality systems use three basic tracking mechanisms to provide a high level of engagement for the VR users with the environment [5]. These include:

i. *Head Tracking*

Head tracking refers to the six degrees of freedom (6DoF) that plots the movement of the users head over three axes. The movement includes pitch, yaw and roll. Head tracking is achieved through the deployment of the three sensors mentioned in the previous section. It is important to note that the head tracking systems are sensitive to latency and values higher than 50ms will cause a perceptible delay in head movement and the response in the virtual environment.

ii. *Eye Tracking*

Eye tracking involves the tracking of the human eye movements with the virtual environment with a view to providing depth of field and generate a realistic experience. The tracking is done with the use of infrared sensors. Eye tracking technologies are yet to mature completely.

iii. *Motion Tracking*

Motion tracking refers to the visualization and interaction of the user with his/her own self. The reality system uses input devices like gloves, controllers and motion platforms along with associated sensors to detect motions and gestures of the user and provide requisite visualizations to the display mechanism (HMD, display screens and/or projection systems) being employed.

iv. *Room-scale Tracking*

Room-scale Tracking is an extension of positional tracking that enables the HMD to track user movements within the limited confines of a fixed space like a room. Room-scale tracking allows the user to remain engaged with the virtual environment while performing a variety of movements like walking, jumping, and running, crawling and kneeling. Room-scale tracking necessitates the placement of additional sensors, generally on the diagonally opposite corners of the space that is part of the virtual environment. Room-scale tracking accords the user complete freedom of movement with the fixed space.

e. **Acoustic Effects**

The sound emanating from a sources travels through air, propagates through material structures, reflects off objects in its path and strikes the ear drum of a user. The human brain processes the inputs received from the ear and estimates various parameters like the distance of the source from the user, the perceived direction of origin, the emotion or other background's associated with the sound and accordingly triggers the appropriate responses. The brain is able to differentiate between the original sound and its reflections and perceive the time lag between them. In order to mimic the real world acoustics, most VR systems employ a 3D stereophonic surround sound system. .

f. **Olfactory Triggers**

Olfactory responses are triggered in a user through the deployment of aromatic diffusers that are based on natural materials and/or chemicals triggered based on manual interventions, scheduled or as per pre-programmed events within the application. The aromatic dispensers are coupled with the visual systems (HMD) and/or audio systems to provide a "truly realistic" "virtual" experience to the user.

g. Processors

It is but obvious that VR applications would require tremendous amounts of processing power. Visualizations require specialized processors and huge amounts of memory to run. The VR system consists of a number of processors handling discrete functions. These include:

i. System Input Processor

The System Input Processor (SIP) is responsible for the control of devices that feed (input) information into the VR system including consoles, position trackers, acoustic inputs etc. The SIP is also responsible for distributing the input information to the other system blocks within specific latency limits.

ii. Simulation Processor

The inputs provided by the SIP are actioned by the Simulation Processor (SP). SP is a core component of the VR system that determines the action corresponding to user inputs.

iii. Rendering Processor

The sensory perceptions in the VR user are as a result of the output from the Rendering Processor (RP). There are separate RP corresponding to each sensory system including visuals, acoustic, olfactory, aromatic and touch among others.

2.5 Key Reality System Parameters

This section presents the key parameters that define the engagement levels or the levels of immersion of a user with the VR system:

1. Field of Vision

The Field of Vision (FOV) also referred to as the field of view is responsible for providing the VR users with a realistic rendering of the real world environment being simulated. FOV refers to the environment observable by the user at any point in time. An increase in the FOV would lead to the enhancement of the user experience. The different values of FOV, for commonly used displays, are illustrated in figure 5 and the common values are presented in table 4.

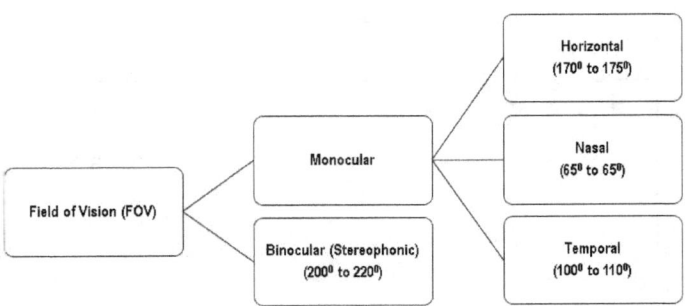

Figure 5 – Understanding the Field of Vision

Most of the commercially available VR Headsets provide a FOV in the range of 90-120⁰.

Table 4 – FOV Values

S.N	Value	System	Remarks
1	90^0	Google Cardboard	-
2	90^0 to 110^0	Windows Mixed Reality	-
3	90^0 to 120^0	Commercially available systems	Standard Range
4	210^0	Start VR/iMAX	Maximum available value

2. Frame Rate

Consecutive images being displayed on the headset are referred to as frames. Frame rate refers to the rate at which the frames are refreshed, creating a perception of continuity or motion. It is important to maintain a constant frame rate so as to maintain the perception of reality. A slow frame rate will affect the user experience as the brain perceives the lag in the action and the expected response. A minimum rate of 60 frames/sec (60 Hz) is required for an immersive VR experience. There are HMD available with frame rates of 90 and rates of 120 would also be available soon

3. **Latency**

Latency refers to the delay in the movement of the head position of a user and the image being displayed on the headset. For realistic experience the latency (motion to photon latency) of the VR systems should be within 20ms. A lower latency increases the user engagement with the virtual environment. On the contrary higher values of latency can result in unnatural lag and result in motion sickness (simulation sickness). This implies that the refresh interval of the associated display systems should be 60 Hz (17ms refresh rate) or 90Hz (11ms refresh rate).

4. **Acoustics**

   The acoustic environment plays an important role in enhancing the levels of user immersion with the VR system. HMD's with integrated headsets as well as those with provision for external headsets are available. Tethered headsets (which require a connection to a personal computer provides a high video and audio resolution. For a truly engaging experience, three dimensional (multi-speaker) audio (3D soundscapes/ 3D audio engine), also referred to as positional audio provides the most realistic audio experience.

3.0 Augmented Reality

Augmented Reality (AR) refers to a real life (physical) environment (or view) that is augmented by introducing (superimposing) computer generated objects thereby altering (enhancing) the users perception of existing reality (real world environment). AR facilitates the augmentation of the real world perception of human beings with the introduction of the unlimited potential of virtual world to enhance our sensory capabilities (vision, sound, smell and sense).

The verb 'augment' can be defined as "to increase the size or value of something by adding something to it". AR can be defined as an augmented view of the real-life environment through the use of virtual objects, with a view of enhancing the reality quotient of the user.

As opposed to VR, Augmented Reality (AR) refers to the enhancement (augmentation) of the real life environment through the introduction of virtual objects that alters the user perception of the real environment. The overlay of the virtual objects and/or information complements/augments the real environment with the aim of achieving specific objectives.

Example 4

A tourist visiting the Taj Mahal, in India can make use of a headset that provides historical information and/or context, as per the user FOV. The actual images viewed by the user would be superimposed with text and/or audio narration to enhance the viewing experience. The process facilitates a higher level of engagement of the user with their immediate surroundings, provides factual and context specific information and enhances the quality of visit.

3.1 Augmented Reality Categories

Augmented Reality (AR) technologies can be categorized based on the application context. The four primary categories include:

1. Marker Based Augmented Reality

Marker-based augmented reality employs visual markers to provide additional inputs to a user. Marker based AR systems provide the augmented view of the environment only when triggered by the user (manual and/or automatic). The simplest method uses a visual marker in the form of a QR/2D code that provides additional information when sensed by the reader (Example – Camera). QR codes provide a means of easy recognition and the advantage of low processing power for detection. The detection of the marker by the sensing mechanism will provided the augmented information.

Marker based AR applications are hard coded and hence simpler than the other categories of AR systems. The AR system also does not require positional sensors, accelerometers or compass.

Example 5

Google Goggles (provides information about a real image – if available in the Google database)

Natural Feature Tracking (NFT) is a special case of marker based AR that uses the real view as a marker. A feature detection mechanism embedded within the application identifies a unique, reproducible feature or point in the real environment, which can be used as a marker (fiducial marker).

2. Markerless Augmented Reality

Markerless Augmented Reality systems uses a location based, position based system (GPS) along with other type of sensors line accelerometer (velocity meter) to provide location based services. Markerless AR systems are popular owing to the availability of smart phones. Markerless AR systems are however more complex, in comparison with marker based applications, due to the need to incorporate pattern recognition and/or image processing algorithms.

Example 6

Direction Maps, ARIS (interactive story telling application)

3. Projection Based Augmented Reality

Projection Based Augmented Reality systems work on the principle of human interaction with projected images (light) in a real life environment. The user interaction (known or unknown) is identified by comparing the projected image with the image altered post the user interaction

Example 7

Virtual Keyboard, 3D hologram projection, Specialized Military Training

4. Superimposition Based Augmented Reality

Superimposition Based Augmented Reality systems create an augmented view of an object, either through partial or full replacement, that is superimposed on the original object in a real environment.

Example 8

An interior decorator can use the superimposition based augmented reality system to propose a new layout or illustrate the effect of the rearrangement of furniture at home or office. Leading furniture retailers provide an image of different furniture combinations that can be superimposed on the real object understand their impact.

3.2 Understanding the Augmented Reality Functional Architecture

Augmented reality applications aim to introduce virtual (computer generated) objects into a real environment. Augmented reality applications facilitate the interaction of virtual objects within a real environment in a manner that is transparent to the user. Augmented reality environments are presented to the user through a self-contained system that eliminates the need for high processing power engines that require tethering support. These includes a variety of display mechanisms ranging from monitors to handheld devices and headsets. The latest offering is the virtual retinal displays that can be compared to contact lenses that present an augmented reality environment to the users, the key system components include:

1. **Cameras and Sensors**

    A combination of sensing technologies and

capturing technologies like camera are crucial components of the reality technologies value chain. A sensor is responsible for capturing the real-world user interactions and provide them to the processing hardware for interpretation and triggering response mechanisms.

Example 9

Depth sensors

The user environment (as per the FOV of the user) is scanned, and data input to the processing engine, by strategically placed cameras. The processing engine formulates a digital model that is presented to the user.

2. **Projection Equipment**

A specific class (category) of Augmented Reality referred to as 'Projection Based Augmented Reality' used miniature projectors to present a user with an interactive environment, using any available surface. The user environment (as per the

FOV of the user) is scanned, and data input to the processing engine, by strategically placed cameras. The processing engine formulates a digital model that is presented to the user.

3. Processing Capabilities

Augmented reality system requires tremendous amounts of processing power. The number of flops required mimics that of the erstwhile mini super computers. In addition, as is obvious, the system requires significant memory along with associated communication hardware, location sensing devices (GPS) and math co-processors in conjunction with sensors like compass, magnetometers, accelerometers and/or gyroscopes.

4. Reflection Surfaces

Augmented reality systems employ an array of mirrors (generally curved), double sides [one reflecting side]) that is used to reflect light to camera and/or user eye (through suitable display mechanisms[1]). The display mechanism usually consist of 'holographic' lenses (that helps in providing a holographic projection to the user). The 'light' processing engine emits two separate streams of light (to both the eyes of the user) through the lens (consisting of separate layers to handle the primary colours of Red, Blue and Green), with varying angles and/or intensities. The result is a holographic, holistic image flashed on to the retina of the system user.

---

[1] Discussed in Section 2

3.3 Control Mechanisms

Augmented reality devices are often controlled either by touch a pad or voice commands. The touch pads are often somewhere on the device that is easily reachable. They work by sensing the pressure changes that occur when a user taps or swipes a specific spot. Voice commands work very similar to the way they do on our smartphones. A tiny microphone on the device will pick up your voice and then a microprocessor will interpret the commands. Voice commands, such as those on the Google Glass augmented reality device, are pre-programmed from a list of commands that you can use. On the Google Glass, nearly all of them start with "OK, Glass," which alerts your glasses that a command is soon to follow. For example, "OK, Glass, take a picture" will send a command to the microprocessor to snap a photo of whatever you're looking at.

4.0 Mixed Reality

The advantages of VR and AR are combined to create a new visualization, of a real environment with dynamic integration of real-life objects, with physical and virtual objects co-existing in real time. Mixed reality (MR) technologies combines the advantages of VR and Augmented reality technologies.

4.1 Augmented Virtuality

Augmented Virtuality is type of mixed reality wherein a real object is introduced into a virtual environment.

Example 10

An advertisement campaign wherein a car is shown driven on a beach through the waves. Rather than getting the car and the associated shooting equipment and crew on the beach the car is shown to interact with the virtual environment representing the beach.

4.2 Mixed Reality Flavours

MR straddles the best of VR and AR technologies to provide a holistic experience to users. It provides users the ability to simultaneously and seamlessly experience a judicious mix of real and virtual environments. Virtual objects moored onto the real environment creating a perception of reality to the user. MR applications include the entire spectrum between 'a fully real' environment and a 'wholly virtual' environment. There are different categories of MR as listed in the following section:

1. **Mixed Environment**

   A mixed MR environment consists of a judicious and seamless mix of the 'real' and 'virtual' environment that is perceived to be reality by the user.

2. **"Virtual" Natural Environment**

The MR system can also present a real environment with a 'virtual' rendering of an environment overlaid on a real environment.

5.0 Telecommunication Infrastructure Challenges

Reality technology applications are bandwidth and latency sensitive and require a robust communication framework. It is expected that there would be over 120 percent (over 60 times the current traffic density) increase in reality technology traffic during the period 2015-2020. The FCC prescribed speed for broadband is 25 Mbps for download and 3 Mbps for upload [6]. The table 5 highlights the bandwidth requirements for immersive augmented/virtual reality applications [7].

Table 5 – Bandwidth Requirements for Common Reality Applications

S.N	Application	Distribution Platform	Bandwidth (Mbps)	Remarks
1	360° Video	HMD	25	Low resolution
2	360° Video	HMD	100	HDTV Resolution
3	Retinal 360° Video	HMD/ Projection	600	4K TV Resolution

The success of the reality technologies and consequently its penetration is highly dependent on the telecom infrastructure quality and availability. The existing telecom architecture would have to evolve to support the high bandwidth and low latency requirements of reality technologies. The table 6 presents a comprehensive list of new age applications (not limited to those based on reality technologies) and their stringent latency and bandwidth requirements. The above table in indicative.

As the need for immersive content increases applications that require higher bandwidth will be developed and deployed.

Table 6 – Latency and Bandwidth requirements of new age applications

Applications	Latency (ms)	Bandwidth (Mbps)
Virtual Reality	1	800
Augmented Reality	1	600
Autonomous Driving[2]	1	10
Tactile Internet[3]	1	1
Video Conferencing	10	100
Gaming (real time)	10	10
Disaster Management	10	0.9
Remote Control (Duplex)[4]	100	10

---

[2] Driverless Vehicle (Car)

[3] New age internet with ultra-low end-to-end latency of 1ms (User Interface 0.3ms + Radio Interface – 0.2ms + wireless node (eNodeB) and core network (0.5ms) and six nine (99.9999%) availability based on cloud based service platforms and virtualized network functions.

[4] Remote control for tactile/haptic (sensory) machines

Remote Device Monitoring	100	1
eCall[5]	100	0.9
Wireless Office (Cloud)	1000	100
Cloud (Personal)	1000	10
Video Streaming	1000	10
Sensor Networks	1000	0.9

Example 11

Each human eye is capable of receiving 720million pixels. Each full colour pixel requires 36 bits and the frame rates are generally 60 fps. This translates to a transmission requirement of around 3 Tbps. The use of compression techniques will reduce the required bandwidth by a factor of 300, which would double in the near future. Assuming a compression factor of 600 the bandwidth required for an immersive experience bordering reality would be in excess of 5 Gbps.

---

[5] Emergency calls from automobiles to emergency services/service centers in case of accidents

The figure 6 illustrates the new technologies that would shape the next generation of telecom networks. The holistic network architecture would need to encompass data centers, especially at the edge, to facilitate distribution of content (low latency) to end-users.

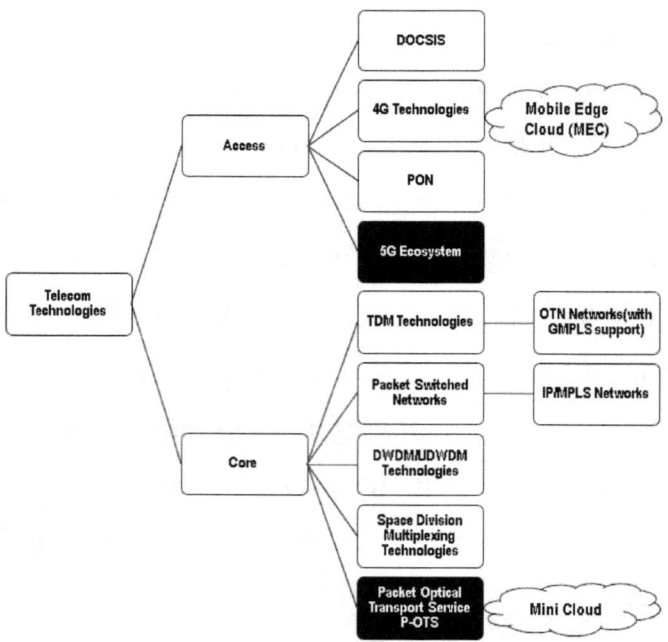

Figure 6 – Telecom Technologies – New Developments

The table 7 summarizes the enhanced requirements of the different network segments, to support high bandwidth, low latency traffic generated by reality technology applications and other emerging technologies like Internet-of-Things (IoT), Tactile Internet.

Table 7 – Structural Changes in Telecom Network to support Reality Technologies

S.N	Layer	Requirements	Impact
1	Backhaul	1. Packet switched network 2. Lean protocol stack / routing engines 3. Use of WDM technologies 4. OTN switching at the network edge	Latency

2	Access	5. Innovative, cost-effective end-to-end design 6. Low power, low maintenance technologies	CAPEX & OPEX
3		1. Enhanced Throughput 2. High QoS 3. Increased storage 4. Reduced latency	Access Architecture and choice of technology, enhanced levels of immersion *(switching at edge, routing at the core, use of DWDM)*
4	Backbone	1. Integration of TDM and Packet switched networks 2. Distributed network architecture 3. Use of Space	Enhanced flexibility, scalability, uptime (six nine to 8 nine) & QoS

		Division Multiplexing Techniques (SDM)	

As highlighted in the table 7, a robust broadband network is a pre-requisite to the successful penetration of virtual reality technologies to the common realm. The high bandwidth and low latency requirements for the reality applications would necessitate a comprehensive architectural review of the telecommunication network – across all layers. The low latency requirements would mandate the deployment/synthesis of data centers, at the network edge, to facilitate the distribution of reality content to the end users. The technologies that are vogue and the ones that are likely to be the drivers in the future are illustrated in figure 6.

The last mile connectivity and throughput needs to be augmented significantly to ensure the delivery of high quality reality content to end users. The combination of optical access technologies and the emerging 5G mobile broadband networks are expected to meet the throughput requirements of the reality applications, minimizing the need for local content storage and provide a fully immersive experience, in the short term. The use of DWDM and Polarization Mode Division Multiplexing (PDM) techniques in the access network coupled with the use of space division multiplexing (SDM) technique in the core is expected to increase available bandwidth significantly while providing for low levels of latency.

However the network needs to account for significant increase in bandwidth as the reality technology matures and achieves mass market penetration. The evolution of the telecom ecosystem to support reality technologies is illustrated in figure 7.

Virtual Reality Telecommunication Systems (VRTS) is the futuristic integrated transmission system that would facilitate the communication of verbal and non-verbal communication over a communication network. The VRTS would facilitate the capturing of the five major sensory perceptions of a user using appropriate sensors, transmit the information to a remote end and replicate the sensory output through an array of actuators.

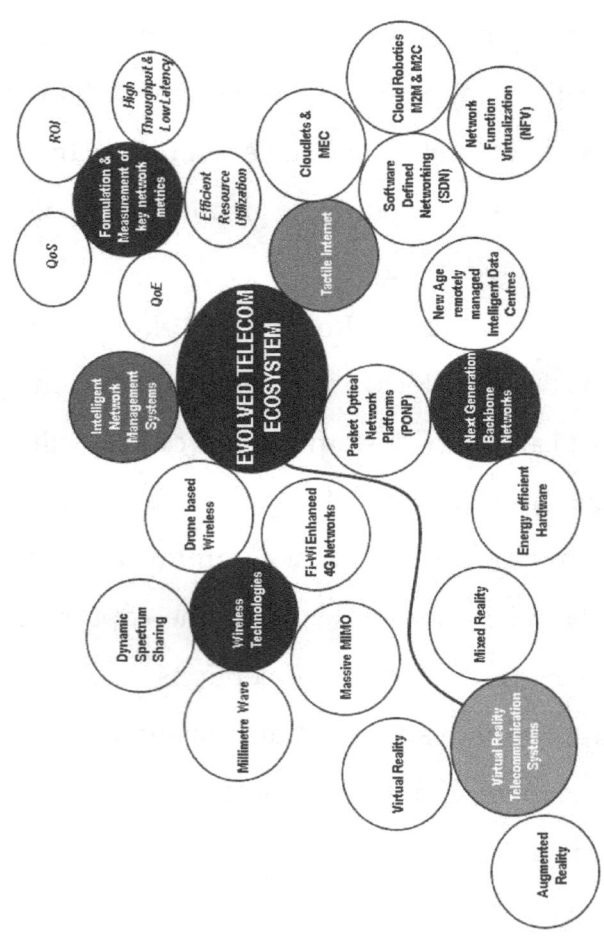

Figure 7 – Evolved Telecom Ecosystem

Referred Standards

1. G.7714/Y.170 - Generalized Automatic Discovery techniques
2. G.7715.1 - ASON routing architecture and requirements for link state protocols
3. G.7715/Y.1706 - Architecture and requirements for routing in automatically switched optical networks
4. G.7716/Y.1707 (01/10) - Architecture of control plane operations
5. G.7717/Y.1708 - Connection Admission Control
6. G.7718/Y.1709 (07/10) - Framework for ASON Management
7. G.798 - Characteristics of optical transport network hierarchy equipment functional blocks
8. G.806 - Characteristics of transport equipment - Description methodology and generic functionality

9. G.8080/Y.1304 (02/12) - ASON Architecture
10. G.8081/Y.1353 (02/12) - Terms and definitions for ASON
11. G.873.1 - Optical transport network (OTN): Linear protection
12. G.983.2 (2002), ONT management and control interface specification for B-PON.
13. G.983.3 (2001), A broadband optical access system with increased service capability by wavelength allocation.
14. G.983.4 (2001), A broadband optical access system with increased service capability using dynamic bandwidth assignment.
15. G.983.5 (2001), A broadband optical access system with enhanced survivability.
16. G.984 - Gigabit-capable passive optical networks (GPON): General characteristics

17. G.984.6 - Gigabit-capable passive optical networks (GPON): Reach extension
18. G.987 – 10 - Gigabit-capable passive optical network (XG-PON) systems
19. G.987.1 - 10 Gigabit-capable passive optical networks (XG-PON): General requirements
20. G.987.2 - 10 Gigabit-capable passive optical networks (XG-PON): Physical media dependent (PMD) layer specification
21. G.987.3 - 10 Gigabit-capable passive optical networks (XG-PON): Transmission convergence layer specification
22. G.989.1 - 40-Gigabit-capable passive optical networks (NG-PON2): General requirements
23. G.989.2 - 40-Gigabit-capable passive optical networks 2 (NG-PON2)

24. G.989.3 - 40-Gigabit-capable passive optical networks (NG-PON2)
25. G983.3 - A broadband optical access system with increased service capability by wavelength allocation
26. G983.4 - A broadband optical access system with increased service capability using dynamic bandwidth assignment
27. P2048.1 Standard for Virtual Reality and Augmented Reality: Device Taxonomy and Definitions (P)
28. P2048.2 Standard for Virtual Reality and Augmented Reality: Immersive Video Taxonomy and Quality Metrics (P)
29. P2048.3 Standard for Virtual Reality and Augmented Reality: Immersive Video File and Stream Formats (P)
30. P2048.4 Standard for Virtual Reality and Augmented Reality: Person Identity (P)

31. P2048.5 Standard for Virtual Reality and Augmented Reality: Environment Safety (P)
32. P2048.6 Standard for Virtual Reality and Augmented Reality: Immersive User Interface (P)
33. P2048.7 Standard for Virtual Reality and Augmented Reality: Map for Virtual Objects in the Real World (P)
34. P2048.8 Standard for Virtual Reality and Augmented Reality: Interoperability between Virtual Objects and the Real World (P)
35. P2048.9 Standard for Virtual Reality and Augmented Reality: Immersive Audio Taxonomy and Quality Metrics (P)
36. RFC 3471 - Generalized Multi-Protocol Label Switching (GMPLS) Signalling Functional Description
37. RFC 3945 - Generalized Multi-Protocol Label Switching (GMPLS) Architecture

38. RFC 4202 - Routing Extensions in Support of Generalized Multi-Protocol Label Switching
39. RFC 4206 - Label Switched Paths (LSP) Hierarchy with Generalized Multi-Protocol Label Switching (GMPLS) Traffic Engineering (TE)
40. RFC 6373 - MPLS Transport Profile (MPLS-TP) Control Plane Framework

Recommended Reading

Books

1. Cedric. F. Lam, *Passive Optical Networks, Principles and Practice*, Academic Press, ISBN: 978-0-12-373853-0
2. Ghein De Luc (2007), *MPLS Fundamentals*.1st Edition, Cisco Press. ISBN-13: 978-1-58705-197-5

3. M. Elanti, S. Gorshe, L. Raman, and W. Grover, *Next Generation Transport Networks – Data, Management, and Control Plane Technologies*, Springer, 2005.

4. S. Kartalopoulos, *Introduction to DWDM Technology*, IEEE Press, Piscataway, HJ, 2000

5. Algirdas Pakštas & Ryoichi Komiya, *Virtual Reality Technologies for Future Telecommunications Systems*, Wiley (2002). ISBN: 978-0-470-84886-9

URL

1. http://www.fpnmag.com
2. www.ftthcouncil.org
3. https://www.broadband-forum.org
4. **http://www.vrstandardsboard.org**
5. http://www.realitytechnologies.com
6. http://www.beyondrealitytech.com

Journals

1. Essiambre, R-J. Et al. "Capacity limits of optical fiber networks." Lightwave Technology, Journal of 28, no. 4 (2010): 662-701.
2. Richardson, D. J., J. M. Fini, and L. E. Nelson. "Space-division multiplexing in optical fibers." Nature Photonics 7.5 (2013): 354-362.

Key Terms

Artificial Intelligence (AI)
Augmented Reality (AR)
Cave Automatic Virtual Environment (CAVE)
Fault, Configuration, Accounting, Performance and Security (FCAPS)
Federal Communications Commission (FCC)
Field of Vision (FOV)
Fully-Immersive VR systems (VR-FI)

Global Positioning System (GPS)

Head Mounted Displays (HMD)

Hertz (Hz)

High-Definition Multimedia Interface (HDMI)

Liquid Crystal Displays (LCD)

Milliseconds (ms)

Mixed reality (MR)

Natural Feature Tracking (NFT)

Organic Light-Emitting Diode (OLED)

Quick Response Codes (QR)

Rendering Processor (RP).

Semi Immersive Virtual Reality (VR-SI)

Simulation Processor (SP).

Six-degrees-of-freedom (6DoF)

System Input Processor (SIP)

Three Dimensional (3D)

Virtual Reality (VR)

Virtual Reality (VR-NI)

Summary

We are the cusp of another major technological revolution that would change the way we live, think and work. There are several major technologies that would usher in a new way of life. This chapter presented the nuances of the reality technologies with specific focus on uncovering the changes in the backbone and access networks. The chapter also introduced emerging technologies like 'tactile internet' with a view of educating the reader on the expectations from the telecom network.

As discussed in this chapter, a robust broadband network is a pre-requisite to the successful penetration and usage of reality technologies by the common man. The high bandwidth and low latency requirements for the reality applications would necessitate a comprehensive architectural review of the telecommunication network encompassing all layers. The low latency requirements would mandate the deployment/synthesis of data centers, at the network edge, to facilitate the distribution of reality content to the end users.

The last mile connectivity and throughput needs to be augmented significantly to ensure the delivery of high quality reality content to end users.

The combination of optical access technologies and the emerging 5G mobile broadband networks are expected to meet the throughput requirements of the reality applications, minimizing the need for local content storage and provide a fully immersive experience.

References

[1] Cambridge Dictionary. (2017). Cambridge Dictionary. Retrieved from http://dictionary.cambridge.org/: http://dictionary.cambridge.org/

[2] Reality Technologies. (2016). Reality Technologies. Retrieved from Reality Technologies: http://www.realitytechnologies.com/

[3] Adams, Ernest (July 9, 2004). "Postmodernism and the Three Types of Immersion". Gamasutra. Retrieved 2007-12-26.

[4] Björk, Staffan; Jussi Holopainen (2004). Patterns in Game Design. Charles River Media. p. 206. ISBN 1-58450-354-8.

[5] Paul, Richard P., Robot Manipulators: Mathematics, Programming, and Control, MIT Press, 1981.

[6] National Telecommunications and Information Administration, National Science Foundation, *THE NATIONAL BROADBAND RESEARCH AGENDA, KEY PRIORITIES FOR BROADBAND RESEARCH AND DATA, Jan 2017. Retrieved from Source URL (https://www.ntia.doc.gov/files/ntia/publications/ national broadbandresearchagenda-jan2017.pdf>*

[7] Mastrangelo Teresa, *Virtual Reality Check: Are Our Networks Ready for VR? June 2016.* Retrieved from: Source URL http://blog.advaoptical.com/virtual-reality-check-are-our-networks-ready-for-vr

CHECK YOUR LEARNING

Review Questions

1. Augmented Reality and Virtual reality refer to the same set of technologies?
    a. True
    b. False
2. Augmented Virtuality is just a different name for Augmented Reality technologies?
    a. True
    b. False
3. The access network bandwidth needs to be scaled to a minimum of _____ to support reality technologies?
    a. 100 Mbps
    b. 1 Gbps
    c. 2.5 Gbps
    d. 10 Gbps

4. In Augmented Reality systems a virtual object is superimposed on a real environment?
    a. True
    b. False
5. In Virtual Reality systems a virtual object is superimposed on a virtual environment?
    a. True
    b. False
6. A high throughput and high latency network is a pre-requisite to the successful deployment of reality technologies?
    a. True
    b. False
7. Virtual Reality technologies require a latency of 1 to 10ms to provide a "fully immersive" experience to the end users?
    a. True
    b. False

8. A _____ HMD requires higher bandwidth for VR application?
   a. Tethered
   b. Streaming Supported
   c. Pixelated
   d. Narrow
9. A mobile broadband network based on the emerging 5G technologies can support reality technologies?
   a. True
   b. False
10. Data centers are required at the network edge to _____?
    a. Distribute content in tune with the latency requirements
    b. Storage only
    c. Support MR applications
    d. Reduce bandwidth requirements

Exercises

1. Write a detailed note on the impact of reality technologies in our daily lives. Provide a comparative table for the different flavours of reality technologies.
2. Prepare a use case for the use of virtual reality technology in the education sector.
3. Enumerate on the specific demands that would be placed on backbone telecom networks due to the emergence of reality technologies.
4. Describe the changes envisaged in the access network/edge to support reality technology applications?

Research Activities

1. What is meant by Virtual Reality Telecommunication Systems (VRTS)?

Briefly describe its hardware and software architecture.
2. Describe the architecture of access networks to support the emergence of reality technologies?
3. Provide a detailed note on the use of DWDM technologies for supporting reality technologies.
4. List the features and specifications of Tactile Internet.

About the Author

Sudhir Warier has over 23 years of corporate leadership exposure spanning the telecom and IT industries. He has had the opportunity to be associated with the entire Telecom & IT value chain, both in the technology and management space. He is a Fellow of the Institution of Electronics & Telecommunication Engineers (IETE), Chartered Engineer (IETE), Graduate in Electronics & Telecommunication Engineering (BE), Post Graduate in Financial Management (MFM), Master of Philosophy – Management (M.Phil).

He is a member Board of Studies, Advisor to some of the leading management and engineering schools in India, is on the external examiner panel of the University of Mumbai, Reviewer for leading technology and management journals and has authored/self-published 12 books and over 30 research papers. His book on knowledge management is a reference text for several national and international universities for diverse fields ranging from management, engineering to science.

www.ingramcontent.com/pod-product-compliance
Lightning Source LLC
Chambersburg PA
CBHW050233230526
45470CB00005B/1929